I0396400

THIS BOOK BELONGS TO:

BLAST OFF!

www.ingramcontent.com/pod-product-compliance
Lightning Source LLC
Chambersburg PA
CBHW080939170526
45158CB00008B/2313
* 9 7 8 1 0 9 7 1 6 8 8 2 8 *